• STAY SAFE • STAY S
• STAY SECURE • STA
• THINK AHEAD • TH

GW00707903

"TRAVEL"
THE WORLD
WITH
"CONFIDENCE"

STAY SAFE, STAY SECURE, STAY SAFE, STAY SECURE, STAY SAFE

THINK & PLAN
AHEAD

Published by 57th Parallel Publications

ACKNOWLEDGEMENTS

The author wishes to thank the following individuals for their help and advice during composition and prior to publication. Sally Stewart, who put the Health information together and gave great assistance. Helen Gall, Douglas Prosser, James and Moyra Robertson and Kate Pedelty who read my material and made various suggestions. Mike Robertson of P. Scrogie, Peterhead who gave me invaluable advice and my Chinese friend, All China journalist, Ruby Zhu Yu who contributed an aphorism of his country. Also, to my family and dear, late mother whose support and encouragement made me take the decision to go ahead with the publication.

CARTOONS were executed by professional artist, Stephen B. Whatley, P.O. Box 2556, Enfield, London EN2 8SE.

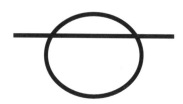

This Booklet has been published by 57th Parallel Publications.
First published in Great Britain 1998. First written 1995.

ISBN 0 9532852 0 0

DISTRIBUTION and typesetting by P. Scrogie, Printers and Publishers, 17 Chapel Street, Peterhead AB42 1TH.
Sales contact: Tel. No. (01779) 476373 Fax. No. (01779) 470003

Printed and bound in Great Britain by:
Caledonian International Book Manufacturing, Bishopbriggs, Glasgow.

<u>CONTENTS</u>   *Page*

INTRODUCTION

I have written this booklet as a concise aid to the traveller abroad. It's kept as short as possible to keep it precise, reasonably priced and to fit one's pocket for ease of reference.

My credentials for my readers information are: I am fifty-eight years old; have travelled extensively in well over one hundred and twenty countries and territories in all five continents over the past forty-two years. I started to travel at the ripe old age of fifteen and hitch-hiked where I could. By the age of seventeen, I was hitch-hiking through Europe, often sleeping rough under the stars and "riding-the-rods" under the freight wagons as the fastest "free" method of covering long distances. This last method is no longer feasible nor legal.

Youth hostels were on the whole, reasonable and cheap; though barns cornfields, park benches, public shelters and on three occasions "enforced accommodation" were alternatives.

The three occasions were – once in Paris for singing in the streets, once in Bruxelles for a pub brawl (one of a crowd of "innocents") and once voluntarily in Chelsea Police Station, London, when I occupied the vacant cell for Gunter Podola, whilst he was on trial for murder at the Old Bailey.

As far as I am aware, I am the last Briton to have worked his passage back from Africa in late 1961 when I joined the National Union of Seamen and signed on as a deck-hand aboard the "Warwick Castle". It was an eventful voyage, the whole crew being dishonourably discharged apart from the officers, myself and a South African who had joined ship at the same time as I did. We docked at King George 5th Docks in London. Thankfully, I now enjoy greater comforts when travelling.

The following statement I have found to be true throughout my life and travels; without it being so, I could never have achieved so much! I have found that approximately 95% of people are honest and decent, with probably another 10% of that percentage being opportunistic with flexible morals and a decidedly dodgy understanding of right and wrong. I have found that this applies to all nationalities and that whatever creed or nationality…all people wish for the same three things…to live in peace with others, to have the right to work for a living and to have the opportunity for education and progress for their family.

I have met many thousands of the ordinary, decent and honest people with the average flaws of humanity and thankfully, I have met very few of that 5% of people who create the worst crime and havoc in life. You cannot advance through this world, life nor travel without placing a certain degree of trust in others. Faith in others honesty has to be accepted at face value.

Thank goodness that the honest folk are such a large percentage!! This booklet is written for and dedicated to that 85% and includes the many fellow travellers throughout the world who have helped me fulfil a fascinating, eventful and generally happy life. During these past forty odd years of travel, I have used every form of transportation, driven in more than sixty different countries and been in all types of accommodation.

During my business career, I have been involved in more than twenty different types of commercial interest, including car and van hire/rental and the Charter Yacht Industry. I retired from business fourteen years ago. To the many friends, advisors and professions I have met during the years and who have all contributed to my education, accumulated knowledge and understanding of our world, I say **thank you.** Over the years of business and travel, I have observed and made personally, many mistakes. This has resulted in the development of many safe, secure and easy methods to improve my own safety, whilst travelling alone in foreign and unknown parts. Most of the time, I could not speak any words of the language of the country and visited the majority of those territories on a solo basis. I have established easy systems for others to follow and use; to gain the one necessity for security.

Confidence!

Speaking the language of the country visited gives one this ability, but following the guidelines and methods outlined in this booklet, will also give you this necessary advantage.

A special thank you to Sally who typed the original manuscript on an electric typewriter that was purchased from a pawnshop in San Antonio, Texas and resold to the same outlet for $2 less, a month later (with new ribbon). Sally also, suggested having the cartoon dividers which were thought-up by myself but ably executed and improved upon by a young, talented, London artist – Stephen B. Whately who has worked internationally as a portrait artist and his work is in many prominent collections. His cartoons are brilliantly executed and the front cover of the book was, from my point of view, nothing short of inspirational.

GENERAL ADVICE

The first lesson of travel to understand is that: **each and everyone of us is a foreigner to all other nationalities**.

Having spent at least one full month of my life waiting in airports, bus and rail stations, ports, taxi ranks and tube stations; the second lesson is **patience**. Be prepared with reading material, crosswords, chess or small computer games. Buy your post-cards early and have them already stamped and addressed for writing and make the best use of your time, there are always post/mail boxes at these places. Relax!! In some countries scheduling is what I'd term flexible.

However, there are times and situations where you will not be satisfied by the instructions, directions or orders given by an official or person in charge. On these occasions, "When in doubt – do nothing", but make yourself obvious. Officialdom will, eventually, do anything and everything to rid themselves of an impending problem or extra paperwork. Nevertheless, at all times be polite and smile as appropriate.

Don't give yourself or others a hard time over tight schedules or connections that can only be met in your country. Always check boarding times, gate or platform numbers and other details of the next step in your journey, at least twice and with more than one official.

There have been many occasions when I have been given the wrong information, including: the wrong station, airport, plane, coach, ship and tickets for the journey!! So, triple check! It saves needless worry, time and sometimes cash.

Remember also, that you are a representative of your own country, and that others will base their opinion of your nation on your behaviour, just as you do about theirs.

Think ahead! Develop good habits!

Get into the habit of keeping documents, tickets, passports and cash in easy to get at places (for you only), and keep them separate i.e. not together in a wallet (unless it is strung around your neck and **under** your clothes – see chapter on money). Nothing irritates a queue of people behind you and the official checking documents, than to wait for you to find them. Another point, that I should like to bring to my readers' attention but will expand on more fully later: "There are no innocents in a crowd or mob". If you see a crowd gathering around an incident or accident, go in the opposite direction. The locals can render assistance more ably and readily than you and you can read the journalists account later…they are paid to investigate. The curious and the "innocent" bystanders are more readily recognised by police forces as blockages or preventing them doing their duty, or worse – silent supporters, not helpful witnesses.

Plan your travels well, make an itinerary of the major towns and cities that you will visit or pass through and check your route on a good, large scale atlas. Read a selection of travel and guidebooks prior to your departure, this investment of time and money will repay you a hundred times over!

Note the sights and the tourist "musts" en route. There's nothing worse than arriving home to be talking to another traveller who's been to the same place as you and finding that you missed a "gem".

Check your health, clothes and visa requirements at least six months ahead of departure time. And don't make tight schedules for yourself that will be hard to meet. Too short stays in cities or areas of interest result in many unhappy, stressed people. I have met many worried, anxious and very unhappy tourists who have literally ruined their holiday of a lifetime, by having spent days altering a schedule which didn't allow for a flooded road, a landslide, a bad air connection or an extra day's free travel on a riverboat! Give yourself time to stop and stare and "smell the roses"!

Always allow for local and public holidays or special festivals. Check ahead and in the guidebooks for details. Bank holidays can be a hungry time without local currency, or an expensive time when confined to changing travellers' cheques or cash at hotels. These captive markets love the lazy, the unwary or the forgetful! Costs can accelerate at an alarming speed and it's during these times that Murphy's Law is activated – "if more than one thing can go wrong at the same time, requiring extra expenditure, it will!"

At this point, I should like to give my readers the benefit of:

"The Traveller's Ten Commandments", written by an anonymous American and copied from the hand-written, original text which is displayed on the wall of the Hotel de Los Angeles, Oaxaca, Mexico.

The Traveler's Ten Commandments

(American spelling of traveller)

☆ Thou shalt not expect to find things as thou has them at home, for thou left thy home to find things different.

☆ Thou shalt not take anything too seriously, for a carefree mind is the beginning of a fine vacation.

☆ Thou shalt not let the other tourists get on thy nerves, for thou art paying out good money to have a good time.

☆ Remember thy passport, so you know where it is at all times; for a man without a passport is a man without a country.

☆ Remember to take only one half the clothes you think you need – and twice the amount of money.

☆ Remember if we were expected to stay in one place, we would have been created with roots.

☆ Thou shalt not worry, he that worrieth hath no pleasure; few things in life are ever fatal.

☆ Thou shalt not judge the people of a country by the one person with whom you've had trouble.

☆ Thou shalt not make yourself too obviously (American) – when in Rome do somewhat as the Romans do.

☆ Remember, thou art a guest in every land – and he that treateth his host with respect, shall be treated as an honored guest.

Anon

The brackets around "American" are mine, as I feel that any nationality could be inserted though the author was intending his remarks for fellow Americans. In my opinion, it sums up the spirit of travel. Anyone who finds fault with these concepts should stick to the five-star, luxury hotels and organised tours of limited duration; they are not cut out for travel.

Organise your system of payments well. Use whichever method you feel most comfortable with. (Personally, I use cash, using travellers' cheques for emergencies only and have seldom used credit cards.) Bank to Bank wire transfers are generally reliable and completely secure, and from my point of view, less of a worry. (See the money section.)

Don't make yourself a target for the criminal. Whilst the professional criminal can easily identify the "tourist" in a crowd, don't make it easy for him. Flamboyance in any form is an invitation; too much jewellery, flash or loud clothes, cameras strung everywhere, a lot of luggage, reading brochures or street maps whilst looking puzzled or lost. Check with the populace on the street, if the men aren't wearing ties or the ladies not wearing scarves – remove yours! Blend with your environment.

Talking loudly in your own foreign language draws the attention of the thief to the presence of a stranger.

Be aware! Think ahead! Review everything! Think before you act!!

Whilst it is not always possible to do all these things at all times, if you have developed good, safe habits, you will react instinctively and correctly in difficult or dangerous situations.

We are all capable of inattention, day-dreaming, being involved in deep conversation and/or gazing at amazing sights, accidents or the unusual.

That is when the professional thief or thieves, strike!!

HEALTH

It is strongly advised to seek your own doctor's advice, prior to visiting any visiting foreign countries, especially those with tropical climates. You will then obtain the best up-to-date information, on the required and advised necessary immunisations.

The World Health Organisation (W.H.O.) recommends that all travellers now carry a Sterile Medical Pack which has the following items included: syringes, needles, infusion set and suture materials. This equipment is for use by medical personnel and ensures that you have sterilised equipment with you, if needed.

Remember to take an adequate supply of any regular medication you require and in the case of prescribed drugs, a letter of authority from your doctor. It is also recommended that you visit your dentist before leaving.

Basic health rules become very important in hot, humid climates. Greater attention and care must be given to simple cleanliness. Always wash hands prior to eating and after visiting the lavatory. Drink bottled fluids only and use bottled water for cleaning your teeth, peel all fruit.

Carry information on your person as to having a sterile Medical Pack, your blood-type and any allergy, this can be on a necklace or wrist band.

If bitten by an animal, seek immediate medical attention after washing the wound from a source of clean water. The medical staff will need to know the type and location of the animal and in cases of snakebite, a good description is essential. Have someone inform the authorities of details. There are few problems at known tourist sites where professional guides are employed. Generally, the more you deviate from the beaten path, the greater are the hazards, so greater care and awareness must be taken. Remember also, that your consulate will have a list of recommended doctors, dentists and pharmacies. If you intend to go into deep jungle, bush or mountainous regions, it is recommended that you consult health authorities **six months** before your travels commence. Health recommendations are continually changing but amongst the common problems to prepare for, are:

☆ Sunburn
☆ Altitude sickness
☆ Sea and air sickness

☆ Gastro-intestinal upsets
☆ Insect bites

A small first aid kit for use in treating minor injuries is necessary and can be augmented according to medical advice where necessary. Listed below are the components of a first aid kit which will cover most minor problems.

Medical Kit/First Aid Kit
☆ 1 x 4 inch/10 cm crepe bandage
☆ Small pack of sterile gauze swabs
☆ 1 x roll of 2 cm wide medical tape
☆ Various sizes of Band aids
☆ Small bottle of surgical spirit
☆ Small tub of antiseptic cream
☆ Various sizes of foot cushion pads
☆ Scissors
☆ 2/3 larger sizes of waterproof adhesive dressings

Pharmaceutical preparations that are useful
☆ Analgesics (Paracetamol or aspirin)
☆ Anti-diarrhoeal agents
☆ Re-hydrating agents (Dioralyte or Rehydrat)
☆ Anti-malarials (if recommended)
☆ Antibiotics (see doctor for advice)
☆ Anti-hystamine ointment and Anti-microbial ointment (for insect bites)

Ancillary List
☆ Extra suntan Lotion
☆ Insect repellant
☆ Ear plugs
☆ Tampons
☆ Condoms or Contraceptive pills
☆ Water sterilising tablets
☆ Indigestion remedies
☆ Foot powder

Your Personal Additions

LUGGAGE

Too many items result in losses!

Think well on the above phrase prior to choosing your luggage. I shall put my own system down and my reasons for such a choice, the reader must decide on its viability and logic.

Take only one suitcase, preferably a hard-shell one with 2/4 or six wheels and a pulling handle or towing strap or both. **Hard plastic** cases are ideal and, if possible, should have both combination and key locks. Internally, it should have dividing straps on both halves and a divider with three zippered pockets for holding faxes or unused/undeveloped films. This kind of suitcase can withstand the rough handling received during travel, is very secure in hotels and protects against the opportunist thief. It is also ideal as a seat when filling in long waiting periods. Wheels make it easily towed. Take the large size, it helps when taking souvenirs to another country for safer/cheaper/ faster methods of despatch home. Take a small Backpack, with carrying handle as well as soft shoulder harness; and if possible, many pockets of varying sizes and good, strong zippers. Whether you have many tiny padlocks is your personal choice. From my point of view, this is not necessary, is confusing to have too many keys, is time-wasting and attracts the wrong type of interest.

However, do not wear it in crowded places – buses are worst – always carry it in these situations.

It is also very helpful if it has a few inside pockets, both zippered and open, also tie-bands with snap-hooks where items can be retrieved quickly and easily.

This backpack/rucksack can act as an overnight bag when visiting off-shore islands or taking a few days tour or short cruise, whilst leaving your main case in secure left-luggage depots. These facilities are provided free in all four or five star hotels but stations and airports provide lockers for which there is a charge.

An extra fold-away nylon bag for beach, pool, picnic, school or college trips is another very useful addition.

Take a shoulder bag (leather lasts better) to contain the immediate daily requirements. If possible, get one where the whole central section opens to maximise content. It should be large enough to carry three or four films, sunglasses and other, pens and notebook, soft-back novel, chocolate, sweets and a small bottle of water and/or fruit. If it has a couple of zippered pockets for documents or tickets (never passport) and undeveloped films or rubbish to be dumped on return to your accommodation; this is an advantage.

Travel as light as you can and be prepared to give clothes or specialist items (one-time use) away because you will gather souvenirs en route.

One warning about your luggage or suitcase, which will travel as cargo on aircraft, trains, coaches and ferries; **never** leave your arrival point on the sayso of an official "that it will be found and delivered to you".* Many aircraft journeys are partially unloaded and continue to other destinations. If your luggage does not come off the aircraft, make an instant fuss, or it will be winging its way to other parts without you! Be polite but insistent!! Mark your baggage for easy identification by you and others who may have to look for it. Use strong, brightly coloured tape or secure straps, use paint, adhesive stickers and label it both inside and out. Warning: Don't put your home address on the outside stickers, it has been known for local thieves to burgle your property at their leisure whilst the owners are on vacation! They get the necessary information at the local airport.

BE AWARE!

* NOTE: This advice is not applicable to the World's major airlines, who will give you a receipt for missing luggage.

CLOTHES

Take as few as possible and buy specialist needs en route. After use, these can then be sold to other tourists or given as presents to helpful guides, hotel staff or other personnel. When leaving an unwanted item in a hotel or making a gift, be sure to write a note stating your intention to dispose of the item/s, sign it and add the date and room number. This will prevent any accusation of theft by management.

Apart from a light sweater or cardigan, it is better to have all your clothes thin and light, with some of them, loose fitting. Then on cold mornings, more layers can be applied and removed as the day warms up. They are also easier to carry, for wearing again as the sun goes down. Whilst still buying thin shirts/blouses, a mixture of short and long sleeved items as a protection from the sun and insects must be borne in mind. A mix of dark and light colours is helpful, worn according to the colour of the terrain on wildlife study trips. Avoid the animal scaring, criminal attracting, bright colours which are only useful when lost at sea. Another advantage of thin clothes is that they are easily washed and dried when facilities for laundry don't exist. **Again**, leave the expensive Gucci, Dior and Armani on the rack, at home. Dress modestly and conservatively and you will be less of a target. The same advice applies to expensive jewellery. **Don't invite crime! You wear expensive items at your peril!** There are many areas where even the cheap watch should be removed. Taking precautions and **thinking ahead** saves a lot of aggravation and money. Take a collapsible umbrella (which will fit into the backpack), small binoculars (size of spectacle case – some excellent ones available), penknife (small enough not to be considered a threat by officials at airports, customs or border crossings), a small torch or flashlight, a quartz travel clock, medical kit (see Health), a sewing kit with scissors and a good toilet bag (leather lasts longer) with compartments or zippered pockets.

Take well-made nail clippers and an electric razor with rechargeable batteries that can operate off all voltages and recharge from all currents. Wet shavers need no advice and don't require an adaptor either. Though ladies might wish an adaptor for their travel irons or coffee/tea mug heater.

A good sturdy pair of walking shoes is advised and never packed; so board all transport wearing them. In addition, a durable, brand-named pair of trainers and a quality, synthetic pair of sandals (salt and water repellant) are the only male requirement (they double as slippers), but ladies may wish to supplement the list. Socks should be light and layered when necessary (I wore four pairs during winter in Tierra del Fuego). A pair of long johns can be life-saving if jeans and trousers are all thin and you visit some countries in their winter periods. Both long trousers and light thin shorts (which double as swimming trunks) are recommended. A small stock of varied size, plastic bags are useful for laundry, damp clothes or other miscellaneous needs. A cagoule or windcheater with good ventilation (they can become unbearably hot), a pair of quality sunglasses and a floppy sun hat which is easily rolled for stowage when not required.

Tip: All pairs of sunglasses and daily-use specs (eye glasses) should have the leg securing screws dipped in Loctite or similar glue; it saves time, expense and annoyance.

Too many pairs of socks, handkerchiefs, ties or shirts are a waste of space. It's better to replace wornout items as you go along.

Your Ancillary List:

MONEY

I have mentioned, briefly under General, the systems of obtaining cash in various countries and though travellers' cheques and credit cards are very well established, they do have drawbacks. I use travellers' cheques for emergencies only and rely on Bank Direct Wire Transfers of cash, which involves sending a Fax with your verifiable signature, the exact address of the receiving Bank and the routeing number, if known. The fax is backed up by a telephone call which is recorded and voice-printed. Many Banks in the United States, Europe, Asia, Middle East and elsewhere, now offer this facility. Whilst the fax and subsequent telephone call may cost up to £30 or $45 for transfers of $5,000, the commission charged is generally negligible or at least, reasonable. Ask first, before nominating a receiving Bank. The transfer usually takes between 36 hours and five working days, depending on the "money game" played. Cards, however, have a limit and a 5% commission rate and you have first to find the appropriate Bank which deals with your type of Credit/Debit Card. And due to the "gremlins" in computer technology, cards are sometimes gobbled-up by the A.T.M. **Think seriously about choice of system!** Sometimes, countries with a closed economy will not issue you with dollars or sterling and convert all funds sent into their local currency. This is easily overcome by planning ahead and taking your requirements with you. Forewarned by the guidebooks is forearmed, adds to the fun and know-how and saves needless worry and expense. **Think and plan ahead!**

Travellers' cheques can have very high commission rates applied by some Banks and Exchanges (I have heard of 25% and been asked for 20%), unless you find the appropriate Bank that deals with your specific kind (usually only in Capital cities). This can take the fun out of travel, so use travellers' cheques for emergency purposes only. But, **don't rely on a single source** of money or you will be at someone's mercy. Cash dollars are acceptable worldwide, though torn (even slightly) and marked or dirty notes may be refused. Credit/Debit Cards have both benefits and drawbacks. Benefits for use at weekends and during holidays, but drawbacks of abuse of carbons (now disappearing) and of possible theft and misuse. They are often subjected to high commission rates, poor exchange rates and can be removed in error. Check, prior to your travels, whether your country's currency is acceptable and easy to change, then choose the method which suits you best and feel most comfortable with.

Travel is an investment that costs money and like all investments, it is better to **spread the risk**. So, protect your money at all times and don't carry it all in the same place, nor all of it in the same form (e.g. T.C.'s, dollars, or local currency) and don't keep all of it on your person. You will find when visiting other countries, that you will require to visit the Capital city on more than one occasion. This can be for a variety of reasons, but most often, because it will be the hub airport with the most connections to other parts of the country. Also, feeder-flights funnel passengers using discounted airfares through this central hub-airport. I suggest that you plan to visit the capital at least twice and if you use bank/wire transfers of cash, you can organise everything accordingly. You will, most definitely, have fewer problems using this system. The larger nations have second or third cities with exactly the same facilities as their capital, these can be used in the same way. This system makes banking, consular registration, mail receipt, visa application and vital supplies much easier, better co-ordinated and greatly enhances your security. If you stay in a particular city for any length of time and use the bank from which to withdraw money, don't be regular with your visits, vary the times and routes. Take a taxi back to your hotel or visit the nearest large hotel restrooms and stow your money away. **Don't take risks! Think and plan ahead!** Work out your intended expenditure for the day and leave the rest in secure holding areas. It's a good discipline to cultivate. Develop good habits.

When carrying your cash for the day's expenditure, split it into three or four wads. Always keep a wad of low denominations with perhaps one or two notes of higher value. Use money from the same pocket or area to pay for petty purchases made on the streets; e.g. coffee, newspapers, sweets, cigarettes etc. The "open" street is the maximum danger for inquisitive and prying eyes. Thieves check on whether it is worth their bother. So again, spread the risk and your cash throughout four pockets or places on your person and only take sufficient for each planned outing. Carry your emergency wad in a doubly secure place. Using a money-belt is excellent, but don't keep all your money in it at all times. It is never a good idea to develop blind faith in any one particular system. Rest assured, it will be tested one day. Better to leave it in the hotel, temporary bank account in the Capital, a safety deposit facility with your airline tickets and passport or, at worst secured in your suitcase when you go sight-seeing. This applies, especially at night or when visiting new, unknown areas. Safety deposits in poor quality two-star hotels and below are not recommended. Carry a photocopy of your passport and a card from your hotel.

The authorities will accept these in or near the town/city where you're staying. **Never** use or carry a wallet! They are too easy to see in jacket/coat or jeans/pants/trouser pockets. They are also, very easy to remove from these areas. Often they are laid down on tables, bar counters, telephone shelves etc and forgotten. My wallet was neatly removed from my inside jacket pocket in South Africa 38 years ago. The thief used a razor blade to remove it from the outside! Always, carry the low denomination wad in the same, using pocket (being right-handed, I use right trouser pocket); so that it is the only wad displayed in public. If you are the victim of a pick-pocket, it's likely to be the only wad that disappears and the only pocket picked. All five other occasions when this happened to me, it was the "using" pocket they went for. Although on one occasion, it was three different pockets and three different hands of pickpockets working as a gang. I was lucky enough on all occasions to escape intact. There are pickpocket teams with worldwide reputations and you'll meet them in any capital city. (See Crime Scams and Danger Areas). Extra pockets stitched into the inside of shirts and trousers can be helpful and a real leather belt with inside zippered pouch, a great advantage (sold in Arab bazaars). Only remove cash from money belts or secret places in the privacy of your room or toilet. **Don't take risks! Think ahead! Stay safe!** There are very useful plastic containers that are completely waterproof, which hang around your neck on a cord. This keeps your spending cash at the beach or pool, dry and intact. Also, there are canvas wallets with a neck-cord to contain passports, money, tickets or T.C.'s. These wallets have great potential as **extra** security. Don't make them your only secure area! A word about beggars. Don't give money to them. Not only does it increase the problem for future tourists, but thieves have been known to watch such transactions, to identify soft, worthwhile targets. Better to give a donation to local charities. The crippled and the blind around churches are a different matter and few criminals are churchgoers. Generally, it is better to ignore all attempts to beg, obtain information from, or to buy things from people in the street. If someone points to a mess on your clothes or shoes and offers to help, ignore them. Walk briskly and resolutely on, saying **NO** firmly. **Stay safe and secure. Develop good habits.**

When walking around towns and cities, wear your shoulder bag on the inside, away from the traffic and in front of you or with your hand resting on it, change shoulders when you change sides of the street. This deters the motor-cycle bag snatchers, common in the Mediterranean countries, although found in others.

Author's Note

I feel that I have to explain what I mean by "money belt" as there are others on the market today, such as "Hip bags" or "Bum bags" which are designed to be worn outside your clothing. The money belts that are most secure, in my opinion, are those which lie flat against your body and are worn under shirt/blouse and trousers or jeans. They do not show through clothing and unless you are strip-searched, they remain hidden and unknown.

The "Hip bags" whilst rendering the wearer some degree of extra security and quick access to documents, have too many drawbacks. They identify you as a tourist and therefore a target. They tempt the wearer to keep all his "eggs in one basket" and like the wallet are easily seen by any potential thief. They are easily removed whilst asleep and by pickpocket gangs, as the straps are quickly and easily cut.

Be aware! Think ahead! Stay safe and secure! Develop good habits!

CAMERAS, Photographic Equipment and all that

Travel as light as you can, although I recognise that there are devoted photographers to whom this advice will be heresy.

However, make it as compact as possible and all within one carrying case which, if a specialist aluminium one, **denotes** what it contains and should be cuffed and chained to the owner. My own method was to carry one of those "idiot proof" cameras with zoom lens, which took perfect photographs in spite of the operator. It was in a leather case which was threaded onto my belt and only at tourist sights did I have it slung around my neck. Use it and put it away, otherwise play it safe. Again, think ahead. Some places will not have the types of film and batteries that you require, so plan your trips with this knowledge in mind.

If you wish specialist equipment and film e.g. underwater cameras etc then **take them with you**. It is very unreliable to buy them at the resorts. A lead carry-bag for 20 or so films is an advantage for airport x-ray machines. Some films, though not all, do get affected and your excellent record of some memorable place is ruined and lost forever. It is worth the trouble in some under-developed countries, making it a habit of getting your films hand-checked. Disposable cameras for underwater or panoramic shots are not recommended to be purchased from anywhere but your own home base. Again, if you are travelling alone, I cannot stress too much the necessity of keeping ostentation to a minimum and of travelling light. Keep your possessions in a compact, organised fashion.

I have come across the budding Steven Speilberg's, festooned like Christmas trees, plus the inevitable cut strap/s which no longer held its expensive item/s.

Finally, a request on behalf of the many interesting tribes and colourful peoples around the world. If photos of a person and his/her possessions are worth taking, then surely it is worth asking permission first and paying a reasonable price. Privacy has a price worldwide. Many of the world's colourful, remote tribes have few items of value and do not make any souvenirs. Their dress, body markings, dance routines and facial adornments, together with their religious rites and fetishes amount to more than 75% of their lifestyle. Invade these areas with empathy and underline all photographic requests with appropriate offers of payment. Your fairness will be appreciated and it can open doors to even greater knowledge and treasures.

Photographs without permission can be construed as theft, and in no way should be taken around military or sensitive Governmental installations.

NOTE: Video cameras become inoperable in high humidity.

ACCOMMODATION

"You get what you pay for" is no longer the truism that it once was. Nowadays, competition is so fierce that it pays to shop around, (especially for stays of more than three nights). To obtain good, discounted rates for medium to long-term stays, see the section on tips to reduce travel costs. For stops of less than three nights duration, it is more difficult to obtain discounts (but not impossible) on quoted terms and timescales play a larger part in the equation. Remember too, that a 20% discount can be worthless if there are no restaurants or cafes near your accommodation or whose restaurant may be vastly overpriced and it is necessary to use taxis or other transport to get down town and see the sights.

The central location can be worth those few dollars more per night. Almost all hotel taxis take advantage of their captive audience. Whilst there is usually a wide range of hotels and other accommodation to choose from in cities of 1/4 million or more, the choice gets radically reduced in the smaller communities or the more outlandish spots. If you are on a tight budget and cockroaches, bugs and doubtful areas of cleanliness don't bother you, then the world is your oyster! Generally, to have good security, good safe food, cleanliness and peace of mind; it is necessary to choose 3 star establishments and above. Although I have had poor food and no hot water in a 5 star-rated hotel. Conversely, I have stayed in some B. & B.'s around the world, which were immaculate.

The poorer the country, the less reliable can be their star-ratings but again, I came across a 4-star hotel in the heart of Buenos Aires which didn't merit two. Unfortunately, I had paid three nights in advance but recovered well by finding a 5 star hotel undergoing refurbishment which reduced their price to less than the 4-star because of the smell of paint. Joss sticks remedied that. An essential for travel is a good guidebook but remember guide books are just that – guides – they are not infallible. Hotels and restaurants change hands and change management with great regularity, so don't complain when they are wrong. Instead, drop them a P.C. so they can up-date their information, they and the following tourists will be grateful. Generally, you can rely on 98% of the information that an up-to-date guidebook gives. I found the Central and South American Handbooks invaluable during a fifteen month tour of Latin America. Also, great reading at airports or during the journey to the next destination, arriving prepared. It's essential to plan ahead. Even for three nights, it pays to phone about accommodation from airports, bus or rail stations. The guidebooks give the relevant telephone numbers and addresses. At many airports, there are courtesy free-phones where cut-rates are given to the traveller. Even when intending a long-stay **never** book more than the first three nights, it gives you flexibility. Nine times out of ten, you can extend your stay if your accommodation pleases and it can often save you cash and discomfort. Package tours can be a different kettle of fish, however, on the whole, if purchased from a reputable company, are usually good and great value for money. I tend to buy any package tours from the neighbouring country's travel agents as I go along. That way there are fewer problems, better value and more up-to-the-minute information on changes to tours and off-season discounts etc. At least with package tours, all the varied connections are their problem but you do waste more time waiting at airports etc. Having caught many trains, aircraft, ferries and coaches with literally minutes to spare whilst free-lancing, it makes the "two hours prior to departure" requirement by tour companies appear unnecessary. You shave their **time** schedules – at your peril!!

During travel to more outlandish spots, jungle or safari trips and mountain climbs; your choice of accommodation is necessarily limited and on many, it is wise to prepare both mentally and physically for the worst. Give yourself an extra day at the jump-off point to find out if you have forgotten an essential item before it is too late to obtain it. Bottled water, chocolate, torch batteries, flippers and snorkel, mosquito net, insect repellant, long underwear, extra sunscreen etc. Some of these may make all the difference from having an enjoyable experience to having a miserable time. Remember, that jungle, desert and mountain areas have all their stores often physically portered in at much higher costs. Ordinary items like toilet paper and soap may be unavailable or at prices which appear extortionate. In these areas, doing your homework properly and preparing accordingly will alleviate the worst of the rigours to be endured. The mental scars will heal into good dinner-party stories and the physical scars will become trophies that add to the veracity of the tales!

Think and plan ahead!!

FOOD AND DRINK

As a general rule, in many countries of the world, it is safer to drink only bottled water and to take only bottled beers and soft drinks. Too many water supplies have amoeba and bacteria in them and are untreated. It is unwise to drink from streams in the countryside. Whilst the local people may build up a natural immunity to their own water supplies, they can have a severe effect on the traveller. The colourful names of Delhi Belly, Inca Two Step, Gyppy Tummy or Montezuma's Revenge are just a few of the names given to amoebic dysentery found in all hot climates around the world. Tea and coffee made with boiled water are generally safe to drink. In certain areas it is advisable to brush your teeth with bottled water or lemonade, if a clean water supply isn't available.

However, most four or five star hotels have safe treated water, but it is better to ask first. If you travel extensively, rest assured, you will experience the problem personally. Avoid **ice** in drinks, as it's often made from untreated water.

Be careful of salads as they can be washed in untreated water, and be ultra careful of soft fruits (e.g. strawberries), they are often fertilised with untreated sewage. All that can be peeled is safe and can be eaten with confidence but beware of eating too much of it, until your system adjusts.

Properly cooked and hot food is generally safe but beware of the cheaper establishments which may have cracked plates and chipped cups and lower standards of hygiene.

Rare steaks are never recommended outside the luxury class of hotel or restaurant; nor are the very tempting cold dishes of prawns and other seafood. Hamburgers without a brand name chain to back them up are to be avoided, no matter how tempting they look or smell. Chips/French Fries on the other hand have few problems as a purchased snack on the street but meals on trains or at transport cafes require careful thought. When travelling for long periods be prepared with biscuits, chocolate and bottled drinks. Anticipate delays: landslides, floods, earthquakes, fallen trees, impassible bridges, fords and "washed-out" roads. Whilst these are seldom a physical threat to the traveller, they do create problems of delay and cause mayhem with transportation systems. In difficult terrain, mountainous regions, wilderness and jungle; allow for problems in your itinerary.

There are many fascinating spirit drinks around the world which are all safe in moderation. There are also home-made brews which are best avoided if you can do so without giving offence. There are magnificent wines which can compete with and often outshine the known best of the world and cost one twentieth of their price. They make travel most enjoyable! There are too, some great beers, lagers and real ales very much worth trying to your host's appreciation.

Asia, the Pacific, Africa, South America and elsewhere have many unusual delicacies, some with wonderful flavours. Having eaten snake, monkey, horse, camel, goat, dog, guinea pig, kangaroo, ostrich, iguana and alligator knowingly; as well as the considered delicacies of sheeps' eyes, various curds, whey and cheeses, brains, offals and all manner of land and sea snails as well as many types of insect.

I can honestly say that i have experienced a great variety of flavours and food textures, some of them – never again!

However, that's travel and it underlines the truth of the phrase – "One man's meat is another man's poison".

Sometimes it is safer to be circumspect and not overly hungry!

<u>Author's Note</u>

To those who are offended by some of the "delicacies" that I have partaken of in different lands, I can only say that all countries do not enjoy the choice nor abundance of food that the wealthier nations have.

Survival can be the difference between locusts eating your protein or supplying yours in a differing form.

A quote from a Chinese Journalist friend "In China we eat everything that flies, except our aeroplanes and everything that has four legs, except our tables!" Nothing there goes to waste.

ORIENTATION

Other than speaking the language of a country, there is just one major ability to master, that is orientation, which in turn will give the confidence necessary to improve and ensure your security. There are many aids to orientation, the first being a schematic map of the city, town or area with a simple directional cross showing North. If you have chosen a good guidebook, it will have such a plan with the salient points of interest marked on it. If not, your hotel or most accommodation will provide one free. Upon checking in, ask for one and get the management to mark your accommodation on it and also, the potential danger areas. All staff at hotels are helpful in this fashion and the porters will give you much more information for an extra dollar when showing you to your room. They will often provide extra information and brochures throughout your stay and I have even had city tours given by staff members in their off-duty time with their own, personal transport at half the cost of official tours. So, develop this habit of getting the staff on your side. It is always a good idea to take a guided city tour and if you plan to stay for a few days or longer, ask for the tour to be arranged on the second or third day. You will become "au fait" with the immediate area around your accommodation by following a few simple rules. These will really turn you into an instant tour guide, with great knowledge of a new area and how to get around it.

Simple Rules for orientating yourself

When leaving your hotel, complete with street map and card stating the address, telephone number of your hotel (plus names of staff members who speak your language, if necessary). Stand at the entrance for a few minutes and note all immediate landmarks. Are there any spires to be seen and which direction from where you're standing? Tall buildings or differently coloured, oddly shaped or built as opposed to the average? Are there advertising hoardings or neon signs? Look up at the skyline as well as to left and right and both corners of the block. Note all kiosks, vendors, hydrants, mail boxes, traffic lights, telephone or electric connector boxes. These are all good visual and memory stimulating aids for immediate orientation. You will be amazed at how expert your observation becomes and how much more that your memory will retain as you practice the routine. Are there tree-lined avenues or Plazas/Squares and gardens nearby and at which point of the compass to your temporary home? Is it on a bus route and which numbers?

Check which way is north and the time of day. Make a note of the light source. Remember, the sun rises in the east and sets in the west, so the light source will always give your the approximate compass directions, even on a dull day. Get used to this method, cultivate the habit of checking which direction the light source comes from and finding the cardinal points of the compass on your street map. A little practice is all that is required and adjustment made for the time of day. Having established the north and south of your street, or the east and west, and having noted two or three prominent landmarks (write them down if necessary), walk around the block until you return to your starting point. As you gain confidence in yourself, increase this to two or three blocks and make notes. However, a word of warning here. It is better to absorb as much as you can mentally and stop at a cafe or hotel and sit inside or in the lounge area to make your notes and refer to guidebook or maps. **Never** advertise the fact that you're a stranger to the area. **Don't become a target.** Some criminals frequent the street areas around the large, tourist hotels and pick their targets carefully. Those who look confident as they leave expensive areas, and show outwardly that they know where they are going and what they're about, are seldom molested.

If you study guidebooks, maps and tourist brochures on street corners or at pavement cafes, you will invite unwelcome interest and attention. Make a point of entering every four or five star hotel that you encounter on your walks and identify the restrooms, lounges and cafeterias for future use. These hotels are natural "refuges" and secure rest or reference points. They have all facilities under one roof, information, help, telephone, taxis, and someone who speaks your language for translation purposes. Emergencies evaporate or at least become most controllable with access to information and general facilities.

Take stock, think and plan ahead, organise. Stay safe and secure.

CRIME SCAMS AND DANGER AREAS

Most of what is said in this section will be known by most of you but can stand repetition.

Danger areas are unlit back streets, rundown ghetto districts, stadiums, markets, rail and bus stations, metro systems and old cities with narrow streets and multitudes of openings.

Check with your accommodation and request a street map and mark in the danger areas. Use taxis after 9 p.m. and if going out to a night spot in an ill-advised section of the city, check it out in daylight first.

Crime enjoys the anonymity of darkness, so take greater precautions with your own security during the hours of darkness. "If you don't do any intended activity in your own country, then don't attempt it in a strange one" – is a good maxim to hold. This advice applies to activities such as night clubbing, pub-crawling, gambling etc. If you are tempted by the night life then, go with friends, use taxis both ways and only take sufficient cash that you are prepared to spend or lose. Both males and females should be wary of apparently single members of the opposite sex who are overly friendly! Flattery has, always, had a price to be paid. Steer clear of dangerous topics with new-found friends, especially in bars. Politics, religion and sport are three of the better known taboo subjects. Refuse offers of drinks, cigarettes or sweets – **they may be drugged**. Anywhere that crowds gather is the hunting ground of the pickpocket or the gangs. Here is a list of a few of the areas that they operate in:

Sports events	Beauty and tourist spots	Cinemas/Theatres
Airports and all queues	Bus and rail stations	Auctions
Markets	Metros/underground railways	Demonstrations
Underpasses	Fights organised or otherwise	Accidents
Fairgrounds	Parks and around street entertainers	

Beware instant friendship offers!

Beware of the cut-rate, heavily discounted article for sale, just around the corner. Be ultra careful of accepting sweets or food from strangers (see Don'ts). There have been numerous tourists who have had their bank accounts emptied under the influence of a drug similar to scopolamine, administered and given under the guise of friendship. If you suspect that you have become a possible target, make excuses and move elsewhere. **Take instant avoiding action!** Better to be paranoid than mugged. There **are** oddities around and some dangerous oddities, as well as just the dishonest. Try not to attract their attention by looking at them. There is however, an exception to this. Where you feel that you are being carefully observed and closely watched or even followed then, a hard stare or close look can dissuade a potential thief. Remember, if you are accosted, it is better to take **immediate** action. Better to be safe than sorry. With the awkward or the mentally unbalanced, ninety-nine times out of a hundred you can walk away unscathed, with little more than dented pride. The criminal is aware that few tourists will complain to the authorities about theft attempts that failed or when they didn't obtain much. They know that the tourist or traveller has a time-limit and deadlines to meet and they use this knowledge to their advantage. They can spot the "stranger in town" very easily, so don't advertise your first time in that city/town or country. There are those (including some taxi drivers) who will take advantage of it. Being late to catch your plane, train or coach, with just minutes to spare, has resulted in a lot of missing wallets, purses, luggage, tickets and cash. As in the sea, the desert, or in the jungle; signs of distress attract predators! There are too many confidence scams or criminal methods and tricks in use today and they continually change to outwit the unwary. This section, even expanded twenty-fold, will only scratch the surface. It is written for the sole purpose to make the reader **aware** of the problem. **Stay alert in the danger areas!** Chatting loudly in your own foreign language on buses, metros and crowded areas will mark you out as a target for the criminal so, **Be Aware!!**

Finally the system generally used by the pickpocket gangs (although there are numerous variations); they have three, four or five members in the team and select their target carefully. Usually, there is a small diversion created, often in restricted spaces where people are funnelled into narrower areas, like doorways, lifts/elevators, escalators, turnstiles etc. Here one member may stumble or fall, drop a pile of papers, money, tickets or shopping and innocently bend to retrieve it, forcing the person immediately in front of the target backwards and into the target. This creates confusion and imbalance. It is at this precise point, that the pickpocket behind the target goes to work. Five seconds and the whole scam is over. The target is minus wallet, money, purse, tickets and possibly bag and camera straps neatly cut. The booty has been passed to another member of the team, sometimes twice, and it's on its way in another direction. The other team members apply delaying and confusion tactics to cover the operators departure. It's fast and slick! Even if your reactions are equally fast and you catch hold of the thief, the proof is gone (hopefully not) but then, it's your word against his and in another language. And the team will provide back-up.

So, be aware! Think ahead! Take precautions! Try not to become a target. Spread your risk areas! Take immediate action on seeing a possible problem.

Notes of warning

In outlandish parts of some countries, officials cannot be trusted. There have been instances of robbery by policemen and collusion between officials and criminals.

These areas are usually well identified by the guidebooks for the countries that such incidents occur in, and you are strongly recommended to demand that you be taken to the police department or local police station when asked to comply with unusual requests.

Again, in some countries where car crime is common, staged accidents are used to obtain cash and valuables. It is safe not to be out of your car until there is a police presence. Pre-written placards (in the correct language) are advised if touring off the beaten track for long periods (see "Do's").   It is advisable to lock all doors when driving.

Author's Note:

This booklet has been written mainly for the traveller who uses public transportation. While it is wise for all to inform themselves about the Laws of the Country that they visit, it is essential for those who drive, cycle or use private forms of transport. Refer to guidebooks. **Please observe and obey all local legislation.**

I have read and heard about some horror stories concerning organ transplants, slavery, enforced prostitution and controlled lifestyles by criminals. Whilst, some of these have been true in the past, these areas mainly belong to Hollywood fiction and should not deter the intrepid traveller. Kidnapping and terrorism does occur but statistically the chances of it happening to you are infinitesimal. Don't take unnecessary risks and avoid the countries who don't immediately take positive, enduring action. All these things sharpen one's wits and hone the radar, nerve endings. Unfortunately, incidents will continue to happen worldwide, even in the developed nations because of their own ambivalence to the crime. Usually, however, these isolated events happen to the foolhardy, the careless and unwary who spurn good advice.

During the last border conflict between Ecuador and Peru, both sides protected their tourist trade and enabled many tourists to pass from one frontier to the other (c/w Passport stamps) and within a few miles of heavy fighting. Whilst true, this is **not** recommended.

You as a tourist can influence Governments whose methods are undemocratic or anti-human rights. You can refuse to visit them and deny them of your tourist money. Extremist countries or those with dictators should be visited with caution, if at all. Again, on any outbreak of war, go elsewhere.

DO's

☆ Do think, plan and calculate ahead, journeys, costs and requirements. Nothing within your domain, should come as a surprise.

☆ Do develop good habits of forward thinking; and in impending, possibly difficult situations, it pays to "cross your bridges before you come to them".

Whilst worry and anxiety are not recommended:

☆ Do be concerned. Officialdom will otherwise ignore or at least not treat your problem as a priority.

☆ Do be polite at all times. Even when extremely annoyed with the official who cannot see simple, logical facts and truths. Ask politely, to see a more senior official and write down names (only for reference, of course!) It pays to be meticulous when travelling.

☆ Do use money belts and spread your risks.

☆ Do use separate areas for separate important items. Keep your passport, tickets, cash and travellers' cheques separately. It's unwise to "keep all your eggs in one basket".

☆ Do avoid placing temptation in the path of others. Remember, most theft is opportunistic and those with less than you, can succumb. Be compassionate if they do. Recognise your part in the crime.

☆ Do think about the Travellers Ten Commandments.

☆ Do give yourself time to see everything and to plan the next steps. Hurry creates worry and problems.

☆ Do take the city tour and orientate yourself with a map.

☆ Do ask local opinions and advice about danger areas.

☆ Do stay in as good an area as you can afford.

☆ Do eat and drink (when alone) in your accommodation or within five minutes walking distance at night or use taxis both ways.

☆ Do use taxis when visiting tourist sights in dangerous locations and make them wait to return you to your accommodation.

☆ Do walk in well-lit streets at night and ignore all cheap or tempting offers. Nothing is for nothing. There are few bargains to be had on the street.

☆ Do check areas out in daylight. If you must visit the less salubrious areas, use a taxi both ways and go with company.

☆ Do vary your times and routes when visiting your Bank. If you are suspicious about being followed, cross the street or take a taxi. Enter a shop to call one, if necessary and await its arrival.

☆ Do smile and be willing at all times with officials. Remember it's their country.

☆ Do be certain about prices or quotations before agreeing. If necessary, ask twice and have it written down. Carry a notepad and pen for this purpose. (A definite requirement for taxi fares in some countries.)

☆ Do remember your manners. The discourteous oaf in your own country is that bit worse if he's foreign; and everywhere you travel it's **you** who is the foreigner.

☆ Do treat others the way you would like to be treated.

☆ Do register with your consulate, especially if travelling through numerous countries.

☆ Do carry photocopies of your passport and other documents.

☆ Do make use of security boxes where available (not advisable in cheap areas).

☆ Do take precautions every time your leave your room – CHECK!

☆ Do travel as light as possible.

☆ Do check, before travel, on visa requirements, health advice, currency regulations and local laws.

☆ Do know your cancellation procedures for lost or stolen credit/debit cards and travellers cheques. Keep these details separate and destroy any carbons after transactions.

☆ Do have back-up plans for health and other emergencies (e.g. cash).

☆ Do take a list of important phone numbers and addresses with you on your day trips and short cruises etc.

☆ Do inform yourself on exchange rates in advance.

☆ Do remember, that in one or two countries police and other officials cannot be trusted. In some cases of police request, it is safer to comply with them at the police department only. In some countries where car crime against the car driver is common, then:

☆ Do have ready-made placards at hand e.g. Follow me to the police station to report accident/breakdown, or – Please Call Police.

☆ Do make use of 0800 telephone numbers for information to enhance your options especially hotels, airlines and travel agents.

☆ Do remember that Hotel chains can often have 2/3 or more with slight variations of the name in the same city.

Don'ts

☆ Don't wear expensive clothes or jewellery, or make yourself a target. Dress to fit in with the crowd.

☆ Don't carry too many items, or have too many cameras. If you have too much, then someone will relieve you of something.

☆ Don't read maps, guidebooks or tourist brochures in the street. You advertise yourself as a stranger and make yourself a target.

☆ Don't accept sweets, drink, food or cigarettes from casual acquaintances, they may be drugged.

☆ Don't leave luggage unattended for a moment, in any situation, unless under the supervision of an official. If you make certain that your luggage boards the correct flight, train or coach and have the stubs to prove it then –

 ☆ Don't leave the airport, station or bus depot without it.

 ☆ Don't accept that it is lost, without creating a fuss and be prepared to wait a couple of hours, asking polite anxious questions on a regular 15 minute basis. Be concerned but pleasant and thank officials for their efforts even when none is evident.

☆ Don't give money to beggars in the street.

☆ Don't speak to or pat strange dogs. Once you have the smell of an animal upon you, it will attract others. Fleas and skin diseases are less problematical but bites get infected and rabies is often fatal.

☆ Don't investigate movements in the grass or bushes wherever you are, you may find more than you expected. There are a lot of nasties out in the jungle or bush!
When off the beaten track where ants, snakes and scorpions are common –

 ☆ Don't put on clothes and shoes in the morning without checking them out first. In some areas it is advisable to suspend your clothes and backpack from a convenient branch or suchlike. Be cautious.

 ☆ Don't investigate on your own, always get help first! If you look for trouble in a strange country, you will find it.

☆ Don't take things at face value or for granted. Check facts, details and information at the earliest opportunity.

☆ Don't display your cash or expensive items. Use your camera and then put it away again.

☆ Don't complain or shout about other countries methods, systems, habits or time-keeping.

If you get into a difficult situation then –

☆ Don't argue with officials, ask for consular help when necessary.

☆ Don't give your opinion unless asked for it. Especially be careful of taboo subjects e.g. religion, politics etc.

☆ Don't **ever** carry anything for anyone across borders! It may be illegal or contain contraband or drugs and have severe penalties attached!!

☆ Don't take animals, plants, fruits or seeds from one country to another. Disease is easily spread and there are also penalties applied.

☆ Don't take souvenirs of pieces of stone, wood or other integral parts of sites that you visit.

☆ Don't carve or paint your name nor anyone elses to mark your presence at a historic site. Take only photographs and leave only footprints and take rubbish home and dispose of it with integrity.

☆ Don't ever get involved with drugs or dealers

☆ Don't travel overnight on cheap overland buses alone.

☆ Don't have your luggage placed on the roof-racks of buses. Take it inside with you and pay for an extra seat if necessary.

☆ Don't be a hero in the face of an attack.

☆ Don't change money on the street.

☆ Don't follow "helpful" people in the street to find a toilet, shop, bar or inviting bargain that is "just around the corner".

☆ Don't wander aimlessly in strange cities. Always look and be purposeful.

☆ Don't attend sports events, entertainment areas alone.

☆ Don't take more money than needed for the outing or day.

☆ Don't take short cuts at night, stick to well-lit areas.

☆ Don't over-indulge in alcohol or suffer the consequences.

☆ Don't let your suitcase out of your sight when not in secured stowage. Swapping for an empty one of same type, colour and size is worldwide.

☆ Don't swim in strange seas, rivers or lakes, until you have checked with local advice on potential hazards (e.g. Bilharzia, Currents, Crocs., Pollution).

TIPS ON REDUCTION OF COSTS

Almost all methods of cost cutting and saving money require the investment of time. Those in a hurry will, inevitably, pay more for an identical article than one that is carefully researched, but there are many expensive ways of obtaining what you need, that can be avoided.

Most travellers know the methods of using stand-by fares for flights but travel agents have cheap allocations, as well as APEX (Advance Purchase Excursion Fare), on some routes that can reduce fares by up to three-quarters. There are cheap, economy fares worldwide, if you have the time to check and also, to book ahead. Some, though not all, involve travelling at inconvenient hours and perhaps going out of your way via a feeder-flight which may double or triple the duration of the journey. Having sufficient time to travel through the countries chosen, is the key to massive airfare savings. Many countries have 21 or 30 day advance booking, specially discounted, tourist tickets which give the traveller a limited time to tour the country. These have to be utilised sensibly and planned, prior to purchase, as there is seldom sufficient time to see all that you wish to see. It is wiser to make only partial use of these "tourist specials" and utilise them for the very long, and usually most expensive flights. Planning your travels comprehensively saves money. Give yourself time to find things out. Often good discounted fares, tours and accommodation last for a limited period only, as hotels are refurbished and new tours or travel methods are tried out and tested. However, it is always possible to obtain discounts on accommodation for stays of three nights or more (even the 4, 5 star and luxury hotels will discount their rates). You lose nothing by asking for a quotation. If you do, don't allow the porter to load your suitcase on a trolley, prior to establishing the price, and remember to ask, finally, as to whether the price includes breakfast and taxes. You can obtain better deals this way, especially if the hotel has a lot of empty rooms. You can obtain up to 50% discounts for 2/3 week stays and in the best hotels. If you take account of the extras such as TV (CNN News), shampoo, tea and coffee, hairdrier, security and usually central location – such luxury with good discounts makes travel more enjoyable. If your hotel is far away from the centre of things and the available shops, restaurants, cafes and folkloric or other shows – let alone – the tourist sights, you can easily spend the equivalent of another two nights accommodation on travel – taking account of the wasted time as well as the fares. Consider your options. If your time is short, then make use of the professionals – travel agents. Use them from the neighbouring country only, the greater the distance away, the worse the value obtained for your money. The closer the agent is, the more likely that the information has been fully checked out. Also the information will be right up-to-date and the information is less likely to jeopardise his/her reputation with home-based customers. You can stipulate quality and price of accommodation, get instant discounts and seldom be let down. Even after paying through an agent, if you like a hotel after 2/3 nights, it is usually possible to extend and enjoy the same discounted terms. If not, ask for quotes as you come across other suitable accommodation and always ask to see the room. Few hotels will turn away business and give you only the up-front price. (I have stayed in $480 per night rooms for $120 per night, over five nights and enjoyed a $150 per night apartment attached to a five star hotel at $64 per night for 30 days.) Both discounted rates were with full facilities, pool, sauna, exercise room etc. The further down the scale and the busier the establishment is, the less likely is a substantial discount but it's still possible even in high season. **Confidence works wonders.**

A point to be aware of on guided tours where there are two parallel systems in place to accommodate one ethnic group or language area that's different to the host country. If you speak the host country's language well, use it all the time, as "quality hotel allocation" is very definitely biased and all fringe benefits or extras are not evenly divided, including information. When booking a hotel by telephone from the airport or elsewhere and establishing a rate; reserve the right to see the room first, before final acceptance and registration. You will seldom be disappointed and you will save both cash and aggravation. An attic or problem room can be expensive anywhere. The difference in some four or five star hotels, at the same price and in the same city can be "chalk and cheese", so check!! Taxis can be the next major expenditure where costs can be vastly reduced with a little forethought and care. Fortunately, the worst captive market, where the greatest rip-off fares were demanded, namely airports, now have alternatives. Shuttle-buses are reasonably priced and almost available in all countries. However, there is little that one can do when arriving in the small hours of the morning. You might be lucky and get a local bus to a village and get a taxi from there or you decide to pay an exorbitant fare or again, make yourself as comfortable and await normal working hours.

This is an erosion of low-cost, night-flying airfares when balanced against other factors. Never take a taxi in front of expensive hotels. Their rates are nearly always inflated, although I have asked for and obtained reasonable quotations for half and day tours from drivers at the end of the rank. The drivers are always more trustworthy and can often speak your language. Remember to cross to the side of the street which matches your direction before flagging a cab and give the driver a **landmark** address which is close to where you wish to go. It is amazing just how many taxi drivers don't know their city and you can add 1/4 to 1/3 the cost looking for the address. I'm sure that all of you, like me, pay off your cab when stuck in a traffic jam. Watching a meter click over for a half hour is a painful process to this Aberdonian! Finally, make sure that the meter is working and correctly set, or obtain a price for the journey before setting off. Occasionally, in some countries, it's better to dispense with the first taxi that appears when either the vehicle or the driver appears unreliable; and don't let your luggage out of your possession until you are satisfied with everything. Beware of taxi touts. They only increase costs under the guise of helping you to obtain a taxi and carrying your luggage, they're unnecessary. I keep advising fellow travellers to **think ahead.** It is the key to cost reduction and obtaining the best rates. If you are prepared in advance, you don't get caught out. Most times at airports, rail and bus stations, there are banks or cambios/exchanges which are open and available to change currency but not at the best rates, so change only small amounts. However, think of outside normal hours, late flights, holidays, mealtimes and bad connections. Carry a small stock of dollar notes for this type of problem e.g. 5 x $1, 3 x $5, 3 x $10 and 4 x $20 to tide you through these immediate difficulties. Then you can pay porters, taxis or left luggage, have a meal and use the telephone without losing too much through poor rates or by having no facility at all. Most hotels will change dollars or travellers' cheques but at poor rates and sometimes akin to highway robbery. So, be prepared.

On the subject of hotel exchange rates, you can on occasion, make their exchange rate work to your advantage and pay your bill in local currency, having obtained a much better rate at a bank or on the "black market". (Many countries have a black market rate for dollars which is legal and openly used by everyone but be careful, don't make use of this area when alone. Also, in some countries the "black market" trade in currency is illegal with stiff penalties.) Do not contravene another country's Laws.

Don't buy items at hotel shops. You can pay up to 10 times the price of the identical item in the shops in the town or city. This applies to proprietary items, such as toothpaste, batteries, water as well as postcards, films and souvenirs. You have to balance time and convenience with costs, but be aware. Even laundry facilities can be reduced to a tenth of the hotel costs, or at least one fifth. All other facilities and services in hotels are similarly priced at higher rates, so consider your priorities and use your money to its greatest advantage.

When travelling through many countries, one after another, you develop the good habits quickly, think ahead and plan ahead. Visiting capital cities first is a good habit, whether you visit them again is your choice. Plan your tour of each country in a circle, rather than crossing borders when you come to them, unless you're using private transport and then it can pay to cross borders once or twice. However, visas for other countries are much easier to obtain in capital cities and cheaper in time wasted tracking the appropriate embassy down and making your way there. Information and specialised items are easier to find, up-dated film is cheaper, travel agencies have competitors. Consider your requirements before going to tourist sites or areas off the beaten track where you will pay twice the price, or more, for replacements.

I counsel fellow travellers to make the best use of your money, but there are cost-cutting exercises which defeat the object. Such as, low-cost accommodation too far away from the sights that you've come to see! Also, the purchase of tours at the bottom end of the market is often unwise. Invariably, you will be disappointed with these "bargains". Again, appreciation of your guide, if done at an early stage in your tour, works wonders!

When leaving one country for the next, try to work out your last couple of days expenses for extras and don't have too much local currency left to spend or exchange. Sufficient to cover airport taxes (cheaper) and a beer and sandwich. Some countries currency is unacceptable, even in the neighbouring country and cannot be exchanged. Remember the exchanges could be closed for holidays etc.

You have to allow for food and drink, to overcome delayed flights and the use of a porter. However, don't count on using up your local currency in the duty free shops, they may be closed or non-existent or very overpriced. You can be prepared to buy cigarettes (currency for guides, porters if you don't smoke yourself), sweets, chocolate, biscuits, films or spirits to cushion the loss in the exchange rate. It is best to think in dollar terms only when considering duty-free purchases.

Some airlines have high surcharges for luggage over a certain weight, so transfer heavy items, like books, bottles etc to your hand luggage.

Finally, for those with the time to do the necessary paperwork and find the depots for renting cars or other vehicles, you can obtain the latest models at up to half the cost of international car rental companies by making use of the local hire/rental companies.

Author's Note

Having been involved in the business, very successfully for more than twelve years, I know the trade inside out.

If any traveller cannot save a hundred times the cost of this booklet in his/her first month of travel utilising the information contained within, he/she should stick to armchair travel.

COSTS TO BE TAKEN ACCOUNT OF

When planning your travels, allow for the following expenditure:

☆ Air fares, train, boat, coach and taxi fares.
☆ Accommodation – plus taxi/metro/bus fares to and from.
☆ Tips for porters, guides, room service and waiters.
☆ Food and drink and bottled water.
☆ Hairdresser, laundry, shoe repairs and shoe cleaning.
☆ Clothing and other item replacement.
☆ Faxes and telephone calls.
☆ Visas.
☆ Bank commissions.
☆ Airport taxes.
☆ Tour costs, national parks, museums etc.
☆ Entry fees, also taxes levied on same by some countries.
☆ Films and batteries.
☆ Toiletries.
☆ Road Tolls.
☆ Health care – mainly doctor and dentist. You have to pay first, even if insured.

Always allow for problems and for losses. Disasters can happen to expensive items, requiring repairs or replacement. Keep a ticket to get you home or travellers' cheques to the value of an emergency flight home (it could be your health or that of a family member, or an outbreak of war, disease or national disaster within a country where your presence is not helpful).

If travel extends for some considerable length of time, it is helpful and less worrisome for recipients of your postcards to date them and add the day of posting. This makes your travels easier to follow on an atlas which can be enjoyed vicariously by your family and friends.

My own father's sage advice to me when in his mid-seventies – "In strange new buildings of any size, check out the exits and toilets".

Finally a note of great importance. **Have a friend or family member in your home country as an anchor.** This should be someone who knows where you are approximately at all times and that you can depend on fully if a problem does occur.

Safe journey! Bon voyage! Buen Viaje!